Electric Aviation Revolution
Future of Flight

Michael Kante

All rights reserved. No part of this publication may be reproduced, distributed, or transmitted in any form or by any means, including photocopying, recording, or other electronic or mechanical methods, without the prior written permission of the publisher, except in the case of brief quotations embodied in critical reviews and certain other noncommercial uses permitted by copyright law.

Copyright ©Michael Kante, 2023.

TABLE OF CONTENT

CHAPTER ONE   Developing Innovative Battery Technologies for Electric Aircraft

CHAPTER TWO   How do those electric planes handle electric energy

CHAPTER THREE   Charging Infrastructure Innovations

CHAPTER FOUR   Airport design, operations, and grid dependability for Electric Aircraft

CHAPTER FIVE   Hybrid Electric Aircraft

CHAPTER SIX   Urban Air Mobility

CHAPTER SEVEN   Human Factors in Electric Aviation

CHAPTER EIGHT   Electric Aviation in Developing Regions

CHAPTER NINE   Environmental Life Cycle Assessment (LCA) of Electric Aircraft

CHAPTER ONE

Developing Innovative Battery Technologies for Electric Aircraft

When it comes to the development of electric aircraft, the development of better battery technologies is very necessary to meet critical concerns related to range, efficiency, and safety. In this article, we discuss some of the most recent technological developments, such as next-generation fuel cells, high-energy-density lithium-sulfur batteries, and solid-state batteries, and the possible influence that these improvements might have on electric aerial transportation.

1. Batteries that are SolidState
Technology Overview

Structure: Unlike ordinary lithium-ion batteries, which utilize liquid electrolytes, solid-state batteries use solid electrolytes. This basic distinction promotes safety and energy density.

Solid electrolytes may be manufactured from a variety of materials, including ceramics, polymers, or sulfides, each of which has its own set of advantages and difficulties.

Higher Energy Density: Solid-state batteries have the potential to produce energy densities that are two to three times higher than those of lithium-ion batteries now in use, which would allow electric airplanes to go farther.

Safety: The solid electrolyte decreases the possibility of leaks, fires, and thermal runaways, considerably boosting the safety profile.

These batteries often have higher cycle lifetimes and improved stability over time, which contributes to their relatively longer lifespans.

The Obstacles,
The production of solid-state batteries at a large scale continues to be difficult and costly due to the complexity of the manufacturing process. Material Compatibility: Ensuring compatibility and effective ion transport between the solid electrolyte and electrodes needs more study.

High-Energy-Density Lithium-Sulfur Batteries

Technology Overview

Chemistry: Lithium-sulfur (Li-S) batteries employ sulfur as the cathode material and lithium as the anode. Sulfur is plentiful and affordable, making it an enticing option.

Energy Density: Li-S batteries have the potential to achieve up to five times the energy density of ordinary lithium-ion batteries.

Several benefits
High Energy Density: Theoretically, Li-S batteries can give up to 500 Wh/kg, considerably extending the range of electric aircraft.

Cost-Effectiveness: Sulfur is cheaper and more plentiful than the cobalt and nickel used in lithium-ion batteries.

Environmental Impact: The usage of sulfur, a byproduct of industrial operations, helps decrease waste and is ecologically favorable.

The Obstacles,

Cycle Life: Li-S batteries suffer from short cycle life owing to the polysulfide shuttle effect, which leads to capacity decline.

Stability: Ensuring the stability of the lithium anode and reducing the production of dendrites are major topics of continuing study.

Next-Generation Fuel Cells

Technology Overview

Hydrogen Fuel Cells: Convert hydrogen and oxygen into energy, with water and heat as byproducts. Proton Exchange Membrane (PEM) fuel cells are the most often utilized kind.

Solid Oxide Fuel Cells (SOFCs): Use a solid ceramic electrolyte and operate at higher temperatures, enabling great efficiency and fuel flexibility.

Several benefits

High Efficiency: Fuel cells offer better efficiency relative to internal combustion engines and may be more efficient than existing battery technology in certain applications.

Zero Emissions: Hydrogen fuel cells create only water as a waste, making them an ecologically benign solution.

Extended Range: The high energy density of hydrogen may enable greater ranges for electric airplanes.

The Obstacles

Hydrogen Storage: Storing hydrogen safely and effectively is problematic owing to its low volumetric energy density and high flammability.

Infrastructure: Developing a robust hydrogen refueling infrastructure is important to facilitate broad deployment.

Cost: The high cost of fuel cell devices and hydrogen generation remains a barrier to commercialization.

Potential Impact on Electric Aircraft
Range Improvement:

Solid-State Batteries and Li-S Batteries: Both technologies offer considerable gains in energy density, which may immediately translate to extended flying ranges for electric aircraft.

Fuel Cells: Hydrogen fuel cells have the possibility for substantially greater ranges compared to battery-only systems, particularly for bigger aircraft.

Efficiency Gains

Advanced Batteries: Improved energy densities and efficiency imply that electric aircraft can run more efficiently, needing less frequent charging and decreasing operating expenses.

Fuel Cells: High conversion efficiencies of fuel cells may result in more efficient exploitation of energy sources.

Safety Enhancements:
Solid-State Batteries: Enhanced safety profiles owing to the lack of flammable liquid electrolytes lower the danger of fires and thermal runaway.

Fuel Cells: While hydrogen offers storage issues, fuel cell systems themselves may be constructed to be very safe and dependable.

Environmental Benefits:
Reduced Emissions: All new battery technologies and fuel cells contribute to decreased greenhouse

gas emissions compared to conventional fossil fuels, promoting sustainable aviation.

Material Use: Innovations like lithium-sulfur batteries may lessen dependence on rare and costly elements like cobalt and nickel.

Advanced battery technologies, including solid-state batteries, high-energy-density lithium-sulfur batteries, and next-generation fuel cells, offer the possibility of changing electric aircraft. By solving present constraints in range, efficiency, and safety, these technologies may dramatically boost the feasibility and performance of electric aircraft. Continued research and development, plus investment in manufacturing and infrastructure, will be key in achieving the full promise of these advances for a sustainable aviation future.

CHAPTER TWO

How do those electric planes handle electric energy

Electric airplanes handle electric energy using a mix of efficient energy storage, delivery, and consumption technologies. Here are the essential components and tactics involved:

Energy Storage:
Batteries: The most frequent energy storage method for electric aircraft. High-capacity lithium-ion batteries are frequently employed, offering a balance of energy density, weight, and efficiency.

Supercapacitors: These may supplement batteries by supplying fast bursts of energy, which is important for takeoff and other high-power needs.

Energy Management System (EMS):

Battery Management System (BMS): Monitors and regulates the status of charge, condition of health, and general performance of the batteries to guarantee safety and efficiency.

Power Distribution Units (PDUs): Distribute power from the batteries to different subsystems of the aircraft, such as propulsion, avionics, and environmental controls.

Regenerative Systems:
Regenerative Braking: Converts kinetic energy back into electrical energy during the descent and braking phases, storing it back in the batteries.

Electric Propulsion System:
Electric Motors: Highly efficient motors turn electrical energy into mechanical energy to operate propellers or fans.

Motor Controllers: Manage the power supply to the motors, guaranteeing efficient operation and accurate control.

Thermal Management:
Cooling Systems: Maintain ideal operating temperatures for batteries, motors, and other electrical components to avoid overheating and guarantee efficiency and safety.

Aerodynamic Efficiency:
Lightweight Materials: Use of sophisticated composites and lightweight materials to decrease total aircraft weight, hence increasing energy efficiency.
Optimized Design: Aerodynamic designs minimize drag, requiring less energy for propulsion.

Operational Strategies:

Energy-efficient flying Paths: Optimized routing and flying profiles decrease energy usage.

Energy Monitoring: Continuous monitoring of energy usage enables for real-time modifications and effective management of resources.

Through these components and tactics, electric aircraft may properly manage their energy resources to guarantee safe, efficient, and dependable operation.

CHAPTER THREE

Charging Infrastructure Innovations

Wireless charging technology has improved substantially over the last several years, becoming more efficient, adaptable, and widely utilized. Here are some major advances and trends in wireless charging:

1. Improved Efficiency and Power Levels:
Higher Power Transfer: Wireless charging methods have improved to accommodate higher power levels. For instance, Qi wireless chargers currently allow up to 15W for smartphones, while other proprietary systems provide even higher power levels for quicker charging.

Enhanced Efficiency: Advancements in coil design, resonance frequency, and alignment methods have increased the efficiency of wireless charging systems, lowering energy losses and heat production.

2. Expanded Applications:
Consumer Electronics: Beyond smartphones, wireless charging is already ubiquitous in tablets, smartwatches, earphones, and other portable gadgets.
Automotive Sector: Many current electric cars (EVs) are being fitted with wireless charging capabilities, allowing for quick, cable-free charging at home or in public locations.

Medical gadgets: Wireless charging is increasingly employed in medical implants and wearable health gadgets, decreasing the need for invasive procedures to replace or recharge batteries.

3. Technological Innovations:

Resonant Inductive Coupling: This approach enables for more efficient energy transmission over larger distances compared to typical inductive charging, which needs close contact.

RF Energy Harvesting: Radiofrequency (RF) wireless charging may transmit energy over greater distances, but at lower power levels, suited for low-power IoT devices and sensors.

Magnetic Resonance: Allows numerous devices to be charged concurrently and can function over wider distances and around barriers like furniture.

4. Standardization and Interoperability:

Qi Standard: Developed by the Wireless Power Consortium, Qi is the most widely used standard for wireless charging, assuring interoperability across various devices and chargers.

AirFuel Alliance: Focuses on developing resonant and RF-based charging solutions, supporting interoperability and greater use across multiple devices and use cases.

5. Integration into Infrastructure:
Public Spaces: Wireless charging stations are increasingly accessible in public venues like airports, cafés, and hotels, offering convenient charging choices for travelers and commuters.
Furniture Integration: Furniture makers are incorporating wireless charging pads into desks, tables, and worktops, making it easier to charge devices in homes and businesses.

6. Safety and Convenience:
Foreign Object Detection (FOD): Modern wireless chargers contain FOD to identify and prevent the heating of metal items that might possibly pose dangers.

Automatic Power Adjustment: Chargers may adapt the power output depending on the device's needs, guaranteeing safe and efficient charging.

7. Future Prospects:

Long-Range Wireless Charging: Companies are developing technology to enable room-scale or even building-scale wireless charging, possibly enabling devices to charge automatically as long as they are within range.

Dynamic Charging: For electric cars, dynamic wireless charging systems are being researched, which would enable vehicles to charge while in motion, such as while traveling on specially equipped roadways.

8. Environmental Impact:

Reducing E-Waste: By avoiding the need for various charging cords and connectors, wireless charging may help decrease electronic waste.

Sustainability: Continued research into more efficient wireless charging technologies promises to minimize energy usage and environmental effects.

In summary, wireless charging has made great improvements in efficiency, power capability, application breadth, and user comfort. With continued developments and increased use, it is set to become a vital part of daily life across numerous industries.

About aircraft

Wireless charging technology is rapidly being investigated and implemented into the aviation sector, notably for electric aircraft. Here are some particular advances and applications:

1. Wireless Charging for Ground Operations:

Electric Ground Support Equipment (eGSE): Airports are implementing electric ground support equipment including baggage carts and tow tractors. Wireless charging pads installed in airport aprons and hangars allow these vehicles to charge without physical connections, enhancing efficiency and minimizing wear and tear on charging ports.

Electric Plane Charging: Airports are experimenting with wireless charging platforms for electric planes. These pads might be positioned at parking areas or gates, enabling aircraft to charge their batteries during boarding, maintenance, or layovers without the need for burdensome connections.

2. In-Flight Wireless Power Transfer:

In-Cabin Power Solutions: Passengers' rising desire for charging personal gadgets like smartphones, tablets, and laptops during flights has led to the incorporation of wireless charging areas inside seats and armrests.

Wireless Power for In-Flight Systems: Future airplanes may employ wireless power transmission to operate in-flight entertainment systems, lights, and other cabin electronics, simplifying the interior wiring and lowering the overall weight.

3. Maintenance and Efficiency:

Streamlined Maintenance: Wireless charging systems for airplanes might streamline maintenance operations by lowering the amount of mechanical components and connections that need frequent inspection and repair.

Improved Safety and Durability: Wireless methods remove the wear and tear associated with conventional plug-in connections, possibly enhancing the lifetime of both charging infrastructure and aircraft components.

4. Advanced Charging Technologies:

Resonant Inductive Coupling: This technology provides for greater flexibility in alignment between the aircraft and the charging pad, which may be especially advantageous in the dynamic settings of airports.

Magnetic Resonance: Capable of charging many devices at diverse distances, magnetic resonance technology may boost the versatility and convenience of wireless charging for various airport vehicles and possibly for aircraft systems itself.

5. Dynamic Wireless Charging:

Taxiing and Runway Applications: Future advances could allow dynamic wireless charging for airplanes as they taxi to and from runways. This technique would include integrating charging infrastructure inside taxiways, enabling aircraft to recharge without stopping, thereby boosting operating efficiency and decreasing downtime.

6. Environmental and Operational Benefits:

lower Emissions: Wireless charging for electric aircraft and ground support equipment encourages the greater use of electric cars in aviation, contributing to lower greenhouse gas emissions and better air quality near airports.

Noise Reduction: Electric airplanes, assisted by effective wireless charging, contribute to decreased noise pollution compared to conventional combustion-engine aircraft.

7. Challenges and Considerations:

Power Levels: The energy needs of airplanes are substantially greater than those of consumer gadgets, necessitating wireless charging solutions to supply far higher power levels securely and effectively.

Infrastructure Costs: Implementing wireless charging infrastructure at airports demands major expenditure. However, the long-term advantages in terms of maintenance savings and operating efficiency may balance these early expenses.

Safety criteria: Ensuring that wireless charging systems for airplanes fulfill strict safety and reliability criteria is vital. This entails significant testing and certification procedures.

8. Future Prospects:

Integration with Smart Airports: As airports become smarter and more networked, wireless charging systems will be part of a bigger ecosystem that includes autonomous ground vehicles, enhanced energy management systems, and IoT-enabled operations.

Autonomous Aircraft: In the future, autonomous electric aircraft might benefit from wireless charging technologies that offer smooth, hands-free energy replenishment, hence boosting the practicality and efficiency of autonomous operations.

CHAPTER FOUR

Airport design, operations, and grid dependability for Electric Aircraft

The incorporation of wireless charging technology for electric aircraft has various ramifications for airport architecture, operations, and grid dependability. These ramifications include a variety of topics from infrastructure development to energy management and operational efficiency. Here are the major considerations:

Airport Design
Infrastructure Development:
Charging Pads Installation: Airports would need to design and install wireless charging pads at important points like as gates, taxiways, and maintenance facilities. This needs careful design to

guarantee optimum location and interaction with current airport infrastructure.

Space Utilization: Efficient layout design is important to accommodate the additional infrastructure without interrupting present airport operations. This may require remodeling parking aprons, boarding areas, and hangar spaces.

Future-Proofing: Airports must address scalability and future demands, ensuring that the infrastructure can manage growing demand as more electric aircraft enter service.

Safety and Standards:
Regulatory Compliance: Airports must conform to high safety and operating requirements. This includes electromagnetic field (EMF) safety requirements and ensuring that charging devices do not interfere with airplane electronics.

Signage and Guidance: Proper signage and operating instructions will be essential to guarantee that pilots and ground personnel can properly align aircraft with charging pads.

Operations

Operational Efficiency:

Reduced Turnaround Times: Wireless charging may expedite aircraft turnaround times by removing the need for manual plugging and unplugging. This efficiency may lead to shorter layovers and more efficient gate usage.

Automated Systems: Integration with automated and smart systems may further boost operational efficiency. For example, automated docking systems can perfectly line airplanes with charge stations.

Maintenance and Reliability:

Reduced Wear and Tear: Wireless solutions eliminate mechanical wear on charging connections and ports, thus cutting maintenance costs and enhancing dependability.

Simplified Maintenance Routines: Maintenance procedures may be expedited since wireless systems need fewer mechanical components and connections, decreasing the complexity of regular inspections and repairs.

Grid Reliability

Energy Demand Management:

High Power consumption: Electric aircraft and ground support equipment will greatly increase the total power consumption at airports. Managing this demand will need sophisticated energy management technologies to avoid overloading the grid.

Load Balancing: Airports will need to install sophisticated load balancing and energy storage systems to meet peak loads, particularly during heavy traffic times.

Renewable Energy Integration:
Sustainable Energy Sources: To meet rising energy needs responsibly, airports could invest in renewable energy sources such as solar panels and wind turbines. This lowers dependency on the grid and promotes energy resiliency.

Energy Storage Systems: Implementing large-scale battery storage systems may assist manage intermittent renewable energy supplies, offering a reliable power source for wireless charging systems.

Smart Grid Technologies:
Real-Time Monitoring: Smart grid technologies offer real-time monitoring and control of energy use, guaranteeing efficient distribution and reducing energy waste.

Demand Response: Airports may engage in demand response programs, where they alter their energy use in response to grid circumstances, helping to stabilize the system and possibly decreasing energy costs.

Broader Implications

Environmental Impact:

Reduced Emissions: The move to electric aircraft assisted by wireless charging may dramatically cut greenhouse gas emissions and improve air quality near airports.

Noise Reduction: Electric airplanes create less noise compared to conventional jet engines, leading to calmer airport operations and less noise pollution in neighboring towns.

Economic Considerations:

Initial Investment: Significant upfront expenditures will be needed for infrastructure development, including the installation of wireless charging systems and improving grid connections.

Operational Savings: Over time, airports may enjoy financial savings via decreased fuel prices, fewer maintenance charges, and perhaps greater operational efficiency.

The integration of wireless charging technologies into airport operations offers the potential to radically improve airport design, operational efficiency, and grid dependability. While there are hurdles and large expenditures necessary, the long-term advantages in terms of sustainability, efficiency, and lower operating costs make it a hopeful prospect for the future of aviation.

Aerodynamic Design for Electric Aircraft
The move to electric power in aircraft design brings both distinct aerodynamic difficulties and

possibilities. The combination of dispersed propulsion systems, innovative wing designs, and sophisticated aerodynamic control surfaces may boost efficiency and performance. Here's a full analysis of these aspects:

1. Distributed Propulsion Systems

Challenges:
Complex Aerodynamic Interactions: Multiple propulsors spread throughout the wings or fuselage generate complex airflow patterns that must be regulated to minimize undesirable aerodynamic interactions.

Structural Integration: Integrating many smaller engines entails adapting the aircraft structure to accommodate these engines while retaining structural integrity and saving weight.

Thermal Management: Electric motors and batteries create heat, needing effective cooling solutions that do not impair aerodynamic flow.

Opportunities:

Enhanced Lift and Drag Characteristics: Distributed propulsion may enhance lift distribution throughout the wing, possibly lowering induced drag and boosting overall aerodynamic efficiency.

Boundary Layer Control: Strategically positioned propulsors may actively regulate the boundary layer, delaying flow separation and minimizing drag.

Redundancy and Reliability: Multiple engines improve redundancy, boosting safety. If one engine fails, others can compensate, enhancing dependability.

2. Novel Wing Configurations

Challenges:

Structural Complexity: Advanced wing designs such as blended wing bodies or box wings offer structural complications that need to be handled without adding excessive weight.

Aerodynamic Optimization: These combinations need advanced aerodynamic optimization to guarantee they produce the promised performance gains.

Opportunities:
Blended Wing Body (BWB): This design blends the wing and fuselage into a single, seamless structure, lowering drag and increasing lift-to-drag ratios. The BWB design is especially well-suited for electric aircraft because of its potential for effective space management, allowing for greater battery storage spaces.

Box Wing and Joined-Wing Configurations: These designs may minimize generated drag and enhance aerodynamic efficiency. They also provide possibilities for greater load distribution and structural efficiency.

Increased Aspect Ratio: Electric aircraft may benefit from greater aspect ratio wings, which minimize induced drag and boost efficiency during cruise. Advances in materials and structural design help minimize the weight penalty generally associated with longer wings.

3. Advanced Aerodynamic Control Surfaces

Challenges:

Control Complexity: Managing many control surfaces demands complex flight control systems to maintain stability and responsiveness throughout all flight regimes.

Integration with Propulsion: Electric aircraft designs generally need to integrate control surfaces with dispersed propulsion systems, increasing the complexity of the control algorithms.

Opportunities:
Active Flow Control: Technologies like morphing wings and adaptive control surfaces may dynamically alter shape to enhance aerodynamic performance in real-time, enhancing efficiency and lowering drag. Dispersed Control Surfaces: Smaller, dispersed control surfaces may give more accurate control and lessen the aerodynamic penalty associated with larger, conventional control surfaces.

Electric Actuation: Electric aircraft may employ electric actuators instead of hydraulic systems, decreasing weight and improving reaction times for control surfaces.

Integration and Synergy

Optimal location of Propulsors: The location of dispersed propulsion units may be improved to maximize lift and minimize drag. For example, installing propulsors on the wing leading edges may assist sustain connected flow and postpone separation.

Synergy with Structural Design: The aircraft structure must be built to withstand the unusual loading circumstances caused by dispersed propulsion and new wing arrangements. Advances in composite materials and additive manufacturing offer more efficient, lightweight constructions.

Holistic Design Approach: Designing electric aircraft involves a holistic approach that incorporates aerodynamic, structural, and propulsion issues. Multidisciplinary design optimization (MDO) techniques are vital for attaining the highest overall performance.

Future Directions

Hybrid Configurations: Combining electric propulsion with conventional engines (hybrid-electric systems) may assist control range and performance constraints while harnessing the aerodynamic advantages of electric propulsion.

Autonomous Flight Systems: Advanced control surfaces and distributed propulsion systems may be coupled with autonomous flight technologies to boost efficiency and safety.

Energy Harvesting: Some unique designs employ energy harvesting methods, like as solar panels on wings, to enhance the range and durability of electric aircraft.

Conclusion

Electric propulsion brings up new opportunities for aerodynamic design in airplanes. By exploiting dispersed propulsion systems, new wing shapes, and enhanced aerodynamic control surfaces, electric aircraft may make major increases in efficiency, performance, and sustainability. These developments need careful integration and optimization but have the potential to transform the future of aviation.

CHAPTER FIVE

Hybrid Electric Aircraft

Hybrid electric aircraft provide a possible step towards more sustainable flying by combining conventional combustion engines with electric propulsion technologies. This hybrid strategy provides a number of advantages in terms of range extension, fuel economy, and operating flexibility, but also comes with some technological obstacles.

Full analysis of these aspects:

Benefits of Hybrid Electric Aircraft
1. Range Extension
Optimized Energy Employ: Hybrid systems may employ electric propulsion during periods of flight that demand less power (such as taxiing, takeoff, and

landing) and convert to combustion engines for cruising when they are more efficient.

Energy Recovery: Regenerative braking and other energy recovery systems may be utilized to replenish batteries during descent or other low-power stages of flight.

2. Fuel Efficiency

Fewer Fuel Consumption: By depending on electric power for specific stages of the flight, hybrid aircraft may decrease total fuel consumption, resulting to cheaper operating costs and fewer greenhouse gas emissions.

Efficient Engine Operation: Combustion engines may run at their most efficient settings more frequently since the electric system can supply power during times of high demand or poor efficiency.

3. Operational Flexibility

Noise Reduction: Electric propulsion is quieter than conventional engines, which helps minimize noise pollution in and near airports, especially during takeoff and landing.

Redundancy and Safety: Hybrid systems offer various sources of propulsion, boosting safety and redundancy. If one system fails, the other may take over, providing a backup power source.

Versatility in Fuel Sources: Hybrid aircraft may be constructed to utilize a range of fuels, including sustainable aviation fuels (SAFs), further boosting their environmental advantages.

Technical Challenges of Hybrid Powertrains
1. Weight and Space Constraints
Battery Weight: Batteries necessary for electric propulsion add substantial weight to the aircraft.

Balancing this added weight with the requirement to maintain or enhance range and economy is a key design problem.

Space for Components: Integrating both electric and combustion powertrains into the restricted space available on airplanes demands novel technical methods to guarantee all components fit without sacrificing aerodynamics or passenger/cargo room.

2. Energy Management and Distribution

Complex Energy Systems: Managing the allocation of power between electric motors and combustion engines involves complex control systems to guarantee seamless transitions and maximum performance.

Thermal Management: Both battery systems and combustion engines create heat, needing specialized thermal management systems to avoid overheating and assure efficient operation.

3. Reliability and Maintenance

System Integration: Ensuring the smooth integration of electric and combustion systems is challenging and needs extensive engineering to ensure dependability and ease of maintenance.

Maintenance Complexity: Hybrid systems add complexity to aircraft maintenance, requiring specialist expertise and possibly increasing maintenance costs and time.

4. Battery Technology

Energy Density: Current battery technology is restricted by energy density, which constrains the quantity of energy that can be stored compared to the weight of the batteries. Advances in battery technology are required to make hybrid electric aircraft more practical for greater ranges.

Charging Infrastructure: Developing the required charging infrastructure at airports to handle hybrid electric aircraft is a logistical and economical problem.

Potential Configurations for Hybrid Electric Aircraft

Series Hybrid:
Electric-Primary with Engine Generator: In a series hybrid system, the aircraft is predominantly powered by electric motors. A combustion engine functions as a generator to replenish the batteries or supply direct power to the electric motors as required.
Benefit: Simplifies the mechanical connection between the engine and the propellers, possibly lowering weight and mechanical complexity.

Challenge: Efficiency losses in transitioning mechanical energy to electrical energy and back.

Parallel Hybrid:

Combined Power Output: Both the electric motors and the combustion engine may supply propulsion power, either separately or jointly.

Benefit: Allows for variable power management, improving performance and efficiency dependent on the phase of flight.

Challenge: Requires complicated mechanical and electrical integration to handle power distribution properly.

Turboelectric Hybrid:

Electric Propulsion with Turbogenerator: This arrangement employs a gas turbine to create energy, which then drives electric motors that drive the propellers or fans.

Benefit: Can achieve improved efficiency by improving turbine operation and enables for dispersed propulsion.

Challenge: Involves complicated systems integration and needs major developments in electrical systems to manage the power levels needed.

Future Prospects and Innovations

Advanced Materials: The development of lighter and stronger materials may assist offset the weight penalty of hybrid systems, boosting overall aircraft performance.

Battery Improvements: Advances in battery technology, such as solid-state batteries or other high-energy-density options, will be important to make hybrid electric aircraft more feasible and efficient.

Renewable Energy Integration: Incorporating renewable energy sources for ground charging (such as solar or wind) might further lower the carbon footprint of hybrid electric aircraft.

Autonomous and AI-Driven Systems: Enhanced automation and AI-driven control systems may maximize the performance and efficiency of hybrid powertrains, assuring optimum energy utilization throughout the flight.

Hybrid electric aircraft provide considerable potential advantages in terms of range extension, fuel economy, and operational flexibility. However, these advantages come with major technological hurdles that must be addressed by continual research and innovation in areas like as energy management, materials science, and battery technology. As these technologies progress, hybrid electric aircraft might play a vital role in the future of sustainable aviation.

CHAPTER SIX

Urban Air Mobility

Urban Air Mobility (UAM) is an emerging discipline focusing on the development and implementation of air transportation technologies inside urban contexts. This includes the use of electric vertical takeoff and landing (eVTOL) aircraft to offer efficient, quick, and possibly autonomous air transport alternatives for urban and suburban locations. The ramifications for electric aviation are considerable, since UAM intends to cut ground traffic congestion, slash emissions, and alter the way people and products move inside cities.

Design Considerations for eVTOL Aircraft
Aerodynamics and Propulsion:

Lift and Thrust: Efficient lift and thrust systems are necessary for vertical takeoff, hover, and onward flight. This frequently uses a mix of rotors or ducted fans.

Noise Reduction: Minimizing noise pollution is vital for urban areas. Design factors such as quieter rotors and better flight paths are explored.

Energy Efficiency and Battery Technology:
Battery Life: High energy density batteries are necessary to offer appropriate flying range and cargo capacity.

Weight Management: Lightweight materials and structures are important to improve efficiency and performance.

Safety and Redundancy:

Multiple Redundant Systems: To enhance safety, eVTOLs have redundant propulsion and control systems.

Automated Flight Control: Advanced avionics and autonomous flight control systems increase safety and minimize pilot effort.

Human Factors and Comfort:

Passenger Experience: Designing cabins for comfort, convenience of access, and safety measures like emergency exits.

Pilot Interface: User-friendly controls for pilots, with possibility for remote or autonomous operations.

Infrastructure Requirements for Urban Air Transportation Networks

Vertiports and Landing Pads:

Location: Strategically positioned near high-demand locations such as commercial centers, airports, and residential areas.

Design: Includes takeoff and landing zones, charging stations, passenger terminals, and maintenance facilities.

Energy and Charging Infrastructure:

Charging Stations: High-speed charging infrastructure to reduce downtime for eVTOLs.

Energy Supply: Integration with renewable energy sources and smart grid technologies to control energy demand and supply.

Air Traffic Management (ATM):

Urban Air Traffic Control: Systems to handle the growing air traffic inside urban areas, guaranteeing safe and efficient operations.

Integration with Existing Systems: Seamless integration with existing air traffic management systems for coordinated airspace utilization.

Ground Transportation Connectivity:

Multimodal Hubs: Integration with current public transit networks to enable seamless end-to-end travel options.

Regulatory Challenges

Certification and Standards:

Safety Certification: Establishing strict safety standards and certification procedures for eVTOL aircraft.

Operational Standards: Defining operational norms and standards for urban air transportation services.

Airspace Integration:

Low-Altitude Airspace Management: Developing regulations for the management of low-altitude airspace, including defined aircraft corridors and no-fly zones.

Coordination with Authorities: Collaboration with aviation authorities, local governments, and international entities to unify laws.

Noise and Environmental Regulations:

Noise Limits: Setting and enforcing noise pollution rules for eVTOL operations.

Environmental effect: Assessing and reducing the environmental effect of UAM, including emissions and energy usage.

Privacy and Security:

Data Protection: Ensuring the privacy and security of passenger data and operational information.

Cybersecurity: Implementing comprehensive cybersecurity procedures to guard against hacking and other dangers.

Implications for Electric Aviation

Sustainability: UAM may dramatically decrease urban congestion and pollution by converting to electric propulsion.

Innovation: Drives breakthroughs in battery technology, electric propulsion, and autonomous flying technologies.

Market Growth: Opens new markets for air travel, giving potential for new business models and services.

Urban Air Mobility implies a dramatic transformation in how cities handle transportation. The effective integration of eVTOL aircraft into urban areas involves careful consideration of design, infrastructure, and regulatory frameworks to enable safe, efficient, and sustainable operations.

CHAPTER SEVEN

Human Factors in Electric Aviation

The human aspects implications of electric aviation entail ensuring that pilots and crew can successfully and safely operate electric aircraft, particularly eVTOLs. Key concerns include pilot training, cockpit interface design, and the influence of electric propulsion on workload and decision-making.

Pilot Training Requirements
Transition Training:
Differences from Conventional Aircraft: Training programs must cover the differences between conventional and electric aircraft, including propulsion systems, energy management, and flight dynamics.

Simulator Training: Extensive use of flight simulators to create realistic situations for electric aircraft operations, focused on managing electric propulsion systems and emergency procedures.

Energy Management:
Battery Monitoring: Training on monitoring and maintaining battery levels, analyzing energy consumption rates, and maximizing flight efficiency.
Charging Procedures: Instruction on correct charging methods, including the usage of fast-charging infrastructure and safe handling of batteries.

Automation and Autonomy:
Autonomous Systems: Familiarization with autonomous flying systems, including how to observe and intervene in automatic operations.
Human-Autonomy engagement: Emphasis on successful engagement with automated systems,

knowing their limits, and maintaining situational awareness.

Cockpit Interface Design

User-Friendly Displays:

Intuitive Layouts: Cockpit displays must provide information simply and succinctly, with an intuitive arrangement that reduces cognitive burden.

Energy Management Displays: Dedicated interfaces for monitoring battery health, energy use, and remaining flying time.

Enhanced Situational Awareness:

Integrated Navigation Systems: Advanced navigation systems that incorporate urban air traffic data, meteorological information, and energy status.

Augmented Reality (AR): Use of AR to deliver real-time data overlays, boosting situational awareness and decision-making.

Simplified Controls:

Touchscreen Interfaces: Use of touchscreens for simple interface with flying systems, minimizing the need for physical controls.

Haptic Feedback: Incorporating haptic feedback to offer physical reactions to pilot inputs, enhancing control accuracy and feedback.

Impact of Electric Propulsion on Flight Crew Workload and Decision-Making

Reduced Mechanical Complexity:

Fewer Moving Parts: Electric propulsion systems feature fewer mechanical components, lowering the chance of mechanical breakdowns and simplifying maintenance.

Simplified Startup and Shutdown: Electric propulsion provides for speedier and simpler startup and shutdown operations compared to traditional engines.

Energy and Range Management:

Continuous Monitoring: Pilots must continually check energy levels and modify flying settings to optimize battery consumption and prolong range.

Decision-Making under limits: Increased attention on efficient route planning and real-time decision-making to guarantee safe operation under energy limits.

Noise and Vibration:

Reduced Noise Levels: Electric propulsion creates less noise, which helps minimize pilot fatigue and increase communication.

Vibration Reduction: Lower levels of vibration may boost comfort and minimize physical strain on the crew.

CHAPTER EIGHT

Electric Aviation in Developing Regions

Electric aircraft offers the potential to alleviate considerable transportation difficulties in developing nations, especially for distant populations with limited access to traditional transportation infrastructure. By using electric vertical takeoff and landing (eVTOL) aircraft and other electric aviation technology, these locations may boost connectivity, encourage economic growth, and promote environmental sustainability.

Potential Role in Addressing Transportation Challenges
Connectivity and Accessibility:
Rural locations: eVTOLs may provide dependable transportation to rural locations that lack established road or rail infrastructure, enabling crucial access to

medical services, education, and economic prospects.

Disaster Relief: Electric aircraft may play a significant role in providing relief and performing evacuations during natural catastrophes, given to their capacity to fly in varied terrains and circumstances.

Infrastructure Requirements:
Minimal Infrastructure: Unlike conventional aircraft that need substantial airport infrastructure, eVTOLs and smaller electric aircraft may fly from basic vertiports or short airstrips, avoiding the requirement for large-scale infrastructure expenditures.

Off-Grid Operations: Electric aircraft may be supported by off-grid renewable energy sources like as solar panels, which is especially advantageous in locations without stable connection to the power grid.

Socioeconomic Benefits

Economic Development:

Market reach: Improved transportation may allow local manufacturers to reach bigger markets, improving trade and economic activity.

Tourism: Enhanced connection can enhance tourism, creating employment and earning cash for local communities.

Healthcare and Education:

Medical Services: Electric aircraft can give quick access to healthcare institutions and carry medical supplies to rural places, improving health results.

Educational Access: Facilitating the conveyance of educators and educational materials may enhance the quality of education in distant places.

Employment and Skill Development:

Job Creation: The introduction of electric aviation services may produce jobs in operations, maintenance, and administration.

Skill Development: Training programs for pilots, technicians, and other aviation-related positions may boost local skill sets and career possibilities.

Technological Adaptations

Battery Technology:

Durability: Developing batteries that can survive high temperatures and conditions commonly seen in remote places.

Energy Density: High energy density batteries to offer adequate range and payload capacity for operations in broad and sparsely inhabited regions.

Aircraft Design:

Robustness: Designing airplanes that can operate dependably in tough areas and under adverse weather conditions.

Versatility: Multi-purpose aircraft that can be quickly modified for freight transport, passenger services, and emergency missions.

Autonomous Systems:

Pilot Shortage: Implementing autonomous or remotely piloted systems may assist overcome pilot shortages in distant places.

Operational Efficiency: Autonomous systems may boost operational efficiency and safety, lowering the need on local knowledge.

Policy Considerations

Regulatory Framework:

Certification Standards: Developing local certification standards in accordance with worldwide best practices to assure safety and dependability.

Airspace Management: Establishing regulations for the integration of electric aircraft into current airspace, including the assignment of flying routes and operating procedures.

Incentives and Funding:
Subsidies and Grants: Government subsidies and grants may assist the initial investment necessary for electric aviation infrastructure and operations.
Public-Private Partnerships: Encouraging partnerships between governments, private corporations, and international organizations to share expenses and knowledge.

Sustainability Goals:
Environmental laws: Implementing laws that encourage the use of renewable energy and limit the environmental effect of aviation.
Sustainable Development Goals (SDGs): Aligning electric aviation activities with larger SDGs to

ensure that they contribute to sustainable and equitable development.

Electric aircraft shows tremendous potential for tackling transportation difficulties in developing nations, especially in rural villages with limited access to traditional transportation infrastructure. The introduction of electric aircraft may provide several social advantages, including greater connection, economic growth, and enhanced access to healthcare and education. Technological adjustments and supporting legislative frameworks will be required to overcome hurdles and enable the effective integration of electric aircraft into these environments. By harnessing new technology and encouraging cooperation amongst stakeholders, electric aircraft may play a crucial role in advancing sustainable development in underdeveloped countries.

CHAPTER NINE

Environmental Life Cycle Assessment (LCA) of Electric Aircraft

A full Life Cycle Assessment (LCA) of electric aircraft entails examining the environmental implications during the whole lifespan of the aircraft. This encompasses the phases of manufacture, battery production, operation, maintenance, and end-of-life disposal. Here, we highlight the important features and environmental implications connected with each stage:

1. Manufacturing

Materials and Resource Extraction:

Raw Materials: The construction of electric aircraft needs numerous raw materials such as aluminum, titanium, composites, and rare earth elements for motors and electronics.

Environmental Impact: Extraction and processing of these resources may result in substantial environmental implications, including habitat loss, water and air pollution, and greenhouse gas (GHG) emissions.

Aircraft Assembly:

Energy Use: Aircraft manufacturing processes are energy-intensive, requiring activities such as machining, molding, and assembly.

Waste Generation: Manufacturing waste comprises scrap metals, composite trash, and chemical by-products, which need to be handled to reduce environmental consequences.

2. Battery Production

Material Extraction:

Lithium, Cobalt, Nickel: Key ingredients for battery manufacture, which are connected with severe environmental and social implications during extraction.

Mining Impacts: Extraction techniques may lead to soil and water pollution, biodiversity loss, and considerable GHG emissions.

Battery Manufacturing:

Energy Consumption: Battery manufacture is energy-intensive, typically depending on power from non-renewable sources, which adds to GHG emissions.

Chemical Processing: Production includes chemical processes that might release harmful compounds into the environment if not adequately regulated.

3. Operation

Emissions During Flight:

Zero Emissions: Electric aircraft emit zero direct emissions during operation, lowering air pollution and GHG emissions compared to traditional fossil-fuel-powered aircraft.

Noise Reduction: Electric propulsion systems are often quieter, helping to noise pollution reduction.

Energy Source for Charging:

power Generation: The environmental effect of electric aircraft operation is greatly impacted by the source of power utilized for charging. Renewable energy sources (solar, wind, hydro) result in reduced GHG emissions compared to fossil fuels.

Grid Mix: The carbon intensity of the local grid mix effects the total environmental advantages of electric aircraft. Regions with a larger percentage of renewable energy will have a more favorable influence.

4. Maintenance

Maintenance Requirements:

Reduced Maintenance: Electric aircraft have fewer moving parts compared to traditional aircraft, possibly lowering the frequency and environmental effect of maintenance procedures.

Component Replacement: Over time, components such as batteries and electronics may need replacement, which will cause waste and need resource inputs.

5. End-of-Life Disposal

Recycling and Disposal:

Material Recovery: Effective recycling techniques may recover valuable materials from decommissioned airplanes, decreasing the requirement for new raw materials.

Battery Disposal: Battery recycling is crucial to control the environmental effect. Improper disposal may lead to soil and water pollution owing to dangerous compounds.

Waste Management:

Disposal Impacts: Non-recyclable components need to be disposed of in a way that minimizes environmental impact, which may be complex and expensive.

Comprehensive Life Cycle Environmental Impacts

1. Global Warming Potential (GWP):

Operational Phase: Significant decrease in GWP during operation owing to zero emissions.

Manufacturing and Battery Production: High GWP related with energy-intensive operations and raw material extraction.

2. Resource Depletion:

Material Extraction: Intensive resource utilization for materials such as lithium, cobalt, and nickel.

Recycling: Potential to minimize resource depletion via successful recycling systems.

3. Toxicity:

Chemical Use in Manufacturing: Release of harmful compounds during manufacturing and battery manufacture.

Battery Disposal: Risk of harmful leakage from carelessly dumped batteries.

4. Energy Use:

Manufacturing: High energy usage in manufacturing steps.

Renewable Energy for Charging: Transition to renewable energy sources for charging may dramatically decrease total energy demand and GWP.

Conducting a full LCA of electric aircraft demonstrates that although these aircraft have the potential to dramatically decrease environmental effects throughout the operating phase, the advantages are dependant on numerous factors:

Sustainable Manufacturing Practices: Implementing energy-efficient and environmentally friendly manufacturing methods helps alleviate the environmental problems associated with production.

Green Battery Technologies: Investing in research and development of more sustainable battery technologies and effective recycling procedures is vital.

Renewable Energy: Maximizing the use of renewable energy sources for electricity production to charge electric aircraft would boost their environmental advantages.

Policy and Regulation: Strong regulatory frameworks to enable appropriate disposal and recycling of batteries and other components may decrease end-of-life consequences.

In conclusion, electric airplanes provide major environmental benefits, notably in lowering emissions during operation. However, delivering these advantages needs a comprehensive strategy that tackles the full lifetime, from material extraction to end-of-life disposal, stressing sustainable practices and renewable energy integration.

www.ingramcontent.com/pod-product-compliance
Lightning Source LLC
Chambersburg PA
CBHW050236230526
45470CB00005B/1983